NON-DESTRUCTIVE TESTING AND FIELD EVALUATION
OF CHEMICAL PROTECTIVE CLOTHING

Final Report

by

Todd R. Carroll and Arthur D. Schwope
Arthur D. Little, Inc.
Cambridge, Massachusetts 02140-2390

Contract No. EMW-89-C-3045

Project Officer

Robert T. McCarthy
Office Of Firefighter Health And Safety
United States Fire Administration
Federal Emergency Management Agency
Emmitsburg, Maryland 21727

December 1990

ABSTRACT

Chemical protective clothing (CPC) may be contaminated with chemicals during routine and emergency response operations. The chemical contamination may be located on the surface or absorbed in the matrix of the plastic, rubber, or fibrous components of the clothing. This study was undertaken to develop a procedure for assessing the presence of contamination, either before or after decontamination of the CPC. The results from applying the method would aid firefighters in determining the need for and efficacy of decontamination. Furthermore, the results would aid decisions on the reuse of CPC.

The procedure that was developed is based on a volatilization technique utilizing length-of-stain (detector) tubes as the method of detection. The procedure is simple to use, applicable to essentially all protective materials and hundreds of chemicals and mixtures.

The detector tube (DT) technique yields semi-quantitative information on the presence of contamination. Contamination levels are measured as a range of representative concentration (ppm) responses on the detector tubes.

Comparisons can be made among virgin, exposed, and &contaminated material swatches, and among different types of materials. For example, swatches of polyvinyl chloride (PVC) and Chemrel Max® CPC fabrics were exposed to methylene chloride (MeCl) for 15 minutes. After exposure, the swatches were decontaminated, aerated, and analyzed for the presence of MeCl. Swatches of the virgin materials were also tested. No MeCl was measured in either of the virgin materials or the exposed Chemrel Max, however, a response of 700-1,100 ppm of MeCl was detected for the PVC.

The feasibility of the DT technique was demonstrated in the laboratory and the field using more than 15 chemical/material combinations. A technique called Dynamic Thermal Stripping (DTS) was used in the laboratory to validate the DT results. The DT responses were converted to mass (pg) to facilitate comparing DT and DTS test results. Overall, there was good agreement between the amounts of residual chemical measured in the swatches analyzed by DT and DTS. For example, in the case of Viton®/Chlorobutyl and gasoline, 1300-2860 ug of gasoline were measured by DT and 1493 ug by DTS. This overlapping pattern was observed in all except one of the comparative tests.

A simulated field study of the DT technique was conducted with the assistance of the Cambridge (MA) Fire Department. The results were encouraging. The DT technique differentiated the propensity of each of three materials to become and retain contamination. Furthermore, the DT technique could differentiate the level of contamination at different locations on a single garment. For example, the DT responses of ethyl acetate measured in PVC ranged from <200 ppm to >3000 ppm depending on the location tested. Average responses measured at the same relative positions during testing of different materials were 1000-2000 ppm, 1000-2000 ppm, and 200-400 ppm for neoprene, PVC, and Viton/Chlorobutyl respectively.

The field study participants commented that the method was easy to learn and conduct, and provided information that would be of immediate benefit. Furthermore, the participants recommended investigating the applicability of the method to firefighter turnout gear.

This method is applicable to those chemicals for which detector tubes are available. Detector tubes are available for more than 200 chemicals and chemical classes with additional tubes being developed each year. Detector tubes are available from many manufacturers including Dragerwerk AG, Auergesellschaft, Mine Safety Appliance Corp., Sensidyne, SKC Inc., Gilian, Matheson-Kitagawa, etc. While not investigated during this study, other methods of detection (e.g., organic vapor analyzer, pH meters, etc.) are available that could further expand the applicability of this technique.

While not optimized for field use, the method is useful and effective in its present form. All necessary equipment can be either purchased off-the-shelf or easily fabricated. The approximate cost for a DT system composed of three volatilization chambers and associated hardware is $825. On average, the expendable materials required to conduct testing of one garment is estimated at $75.

These costs for assessing the presence of contamination must be added to the price of the garment and the costs of garment decontamination, inspection, maintenance, and storage in any estimate of the total cost of the garment. By dividing these costs among the number of times the garment is used, one can estimate a cost/use for the garment. By comparing the cost/use value for various garment and garment use scenarios, one can develop guidelines for when it is financially advantageous to reuse a single garment multiple times or use multiple new garments a single time.

This study was preliminary in nature and focused primarily on CPC. The procedure requires pre-positioning of protective clothing material swatches on the firefighter's CPC or destruction of the CPC. Both approaches are impractical. It is recommended that an enclosure system be developed that integrates the sample swatch scheme and volatilization chamber to facilitate non-destructive and repetitive testing of garments. In addition, the applicability of the method to other firefighter protective equipment (i.e., turnout gear, gloves, hoods, hoses, etc.) should also be investigated.

CONTENTS

LIST OF TABLES

LIST OF FIGURES

ACKNOWLEDGMENTS

This study was funded by the United States Fire Administration under Federal Emergency Management Agency Contract No. EMW-89-C-3045. We acknowledge the participation and contributions made by Mr. Steve Storment, Deputy Chief of the Phoenix Fire Department Special Operations Branch, as a consultant on this project, and Deputy Chief John O'Donahue, Captain Mike Travers, and Firefighter Ed Friel of the Cambridge, MA, Fire Department for their participation in the field study portion of the project. Additionally, we thank National Drager (Pittsburgh, PA), and specifically Mr. Robert Dusch and Mr. James Fleming, for their generous technical and materials support throughout the program.

INTRODUCTION

Fire department hazardous material response (HazMat) teams are the primary responders to incidents involving hazardous materials in the United States. .HazMat teams often use chemical protective clothing (CPC) during these responses to provide responders with protection from the potential chemical hazards.

CPC is typically fabricated from plastic or rubber materials. There are literally hundreds of CPC materials and an infinite number of chemical/mixtures. The effectiveness of a CPC material or garment as a barrier to a chemical hazard is specific to the chemical/material combination.

Upon exposure of CPC by a chemical, the chemical may be absorbed by the CPC material. In time, any absorbed chemical will diffuse through the material and eventually appear at the inside surface (i.e., breakthrough and permeation). Once there, the chemical becomes available for contact with a wearer's skin and underclothing. Proper selection of the appropriate clothing material will ensure that the resistance of the material to the chemical hazard exceeds the expected duration of exposure.

There are two general classifications of CPC garments: single-use and reusable. A single-use garment is used once and then discarded, a reusable garment is used, decontaminated, inspected, and stored for use at a later date. Cost often times determines whether a garment is considered single-use or reusable.

An important concern in reusing a garment is the efficacy of decontamination following a chemical exposure. Contamination may be of two types: "gross" contamination, and "matrix" contamination. Chemical on the surface of a garment is considered gross contamination. This type of contamination is usually easy to remove and the efficacy of its removal is readily determined visually. Chemical that has been absorbed by a material is considered matrix contamination. Matrix contamination is not readily removed by surface decontamination techniques nor is it readily detected visually or measured.

Matrix contamination represents a potential risk to the next person using a garment. Absorbed chemical will continue to diffuse through the material even after surface decontamination has been completed. Eventually, the chemical will appear at both the inside and outside surfaces of the garment. Matrix contamination presents a risk not only to the next user but also to individuals working in close proximity to garment storage areas as a result of garments outgasing.

In view of the potential and significant hazards associated with matrix contamination, the United States Fire Administration, through the Federal Emergency Management Agency sponsored this preliminary study to develop a method for determining the presence of chemical contamination in protective clothing used by fire department HazMat teams. Ultimately, the objective of the study was to develop a method to be used by HazMat teams to assess the efficacy of decontamination procedures and to aid in reuse decision regarding reusable CPC.

The essence of this project was conceived by the WSFA in the late 1980's when rubbery polymers dominated as the barrier layers of chemical protective clothing. The significant price of total-body encapsulating garments fabricated from these materials required users to decontaminate and reuse the suits to make them cost effective. The late 1980s and early 1990s saw the introduction of a variety of plastic and multi-layered CPC materials that provided improved chemical resistance to a wider range of chemicals. The improved chemical resistance of these newer materials reduces the extent to which they will become contaminated and in turn the extent to which they must be decontaminated. Decontamination, if it is necessary, however, is likely to be more difficult since matrix contamination is more difficult to remove from the newer materials.

Several of the newer materials have been fabricated into relatively inexpensive, limited-use garments which are discarded if exposure to chemical is suspected. This has allowed users to avoid many of the problems associated with decontaminating and reusing clothing. Furthermore, many fire departments are initiating cost recovery programs that requires the organization responsible for the incident to replace the garments and equipment consumed during the response.

These recent developments would seem to limit the need for a means for measuring the presence of chemical contamination in protective clothing used by firefighters. There remains, however, a strong desire to financially optimize the clothing already owned by the fire service and to minimize the clothing that is discarded. Thus, although the situation has changed since this program was conceived, the procedure developed in this study has considerable potential benefit to the fire service.

CONCLUSIONS

Firefighter protective clothing, including chemical protective and turnout gear, that is exposed to chemicals will be contaminated to a greater or lesser extent by those chemicals.

The presence of chemical contamination in protective clothing can be detected by means of length-of-stain (detector) tubes.

A procedure incorporating detector tubes and its feasibility have been demonstrated in the laboratory and the field. In the laboratory, similar amounts of contamination were detected by the detector tube (DT) technique and an instrumental technique, dynamic thermal stripping (DTS) incorporating a flame ionization detector. In the field trial, firefighters from the Cambridge, Massachusetts Fire Department successfully applied the detector tube technique with a minimum amount of training.

In its present form, the procedure is not optimized for field application; however, it is both functional and effective. The procedure requires pre-positioning of swatches of the protective clothing material on the firefighter's outer garment and analysis of the swatches subsequent to chemical exposure or decontamination. While not optimized for field use, the technique is both useful and effective, and could be used in its present form to assess the level and location of chemical contamination in protective clothing. All necessary equipment can be either purchased off-the-shelf or can be easily fabricated in-house.

Performance of the procedure is likely to cost from $100 to $300 or more per garment, depending on whether the identity of the contaminant is known and whether and to what extent the cost of the firefighter's labor is included.

The DT technique could be used as a cost-effective means of saving suits (both CPC and TOG) that would otherwise be considered contaminated.

This technique could be used to reduce disposal costs by aiding in decisions regarding whether or not a contaminated suit can be disposed of as hazardous or nonhazardous waste.

This technique could be easily and successfully conducted by any company and by any individual within a company; however, the participants commented that designating the task to only a few individuals within a company would ensure consistent results.

RECOMMENDATIONS

One or more fire department HazMat teams should test the prototype DT system to evaluate the acceptability and feasibility of the technique under real-life conditions.

The applicability of the DT technique to other firefighter protective equipment (i.e., turnout gear, gloves, hoods, hoses, etc.) should be investigated.

If the DT technique proves successful under actual field conditions, an enclosure system should be developed that integrates the sample swatch scheme and the volatilization chamber. A integrated system would allow for nondestructive and repetitive testing of actual garments.

Alternative detection systems (i.e., HNU®, organic vapor analyzer, etc.) should be investigated to expand the applicability of the technique to other chemicals/mixtures for which no detector tubes (DTs) are available.

Fire departments maintaining significant inventories of reusable CPC should consider applying the life-cycle concepts presented in this report to their specific situations. Identifying the optimum number of uses for their reusable garments will aid in the decisions of when to replace the suits.

METHODOLOGY

OVERVIEW

A simple procedure has been developed to assess the presence of chemical contamination in CPC. The feasibility of the procedure was demonstrated in the laboratory and the field. In the laboratory, three CPC materials were tested against seven chemicals. The results generated using the DT technique were compared to those generated by an analytical laboratory technique. In the field, two fire department hazardous material response team members demonstrated the technique using five protective materials and one chemical. The logistics of the evaluations and the developmental work on the procedure are discussed below.

CONTAMINATION ASSESSMENT CONCEPT

The following criteria were set for an effective method for analyzing protective clothing for the presence of chemical contamination:

- The method must be suitable for use in the field by field personnel,

- The method must be simple, inexpensive (with respect to the cost of using a garment), and require a minimum amount of training to conduct,

- The method must be applicable to volatile and non-volatile chemicals,

- The method must produce results that can be readily interpreted by field personnel.

The contamination assessment concept, which was investigated in this study, is essentially a hot air extraction relying on length-of-stain (detector) tubes as the method for measuring the presence of contamination. A swatch of protective clothing is loaded into a chamber having valved inlet and outlet ports. With the chamber sealed and the valves closed, the container is heated in a hot water bath and the temperature maintained for a specified period. At the end of the volatilization period, the valves are opened and the air inside the chamber is analyzed with a detector tube. The detection of chemical is indicative of the presence of chemical contaminant in the swatch of clothing material.

DTs are widely used for monitoring air quality and indicate the presence and concentration of chemicals by means of color changes. In this application, any color change greater than that of the unexposed, virgin CPC material would suggest the presence of chemical contamination in the swatch.

GARMENT SAMPLING SCHEME AND HARDWARE

Sampling Scheme --

A non-destructive means for measuring the contamination of a chemical protective garment either after an emergency response operation or after the garment had been subjected to decontamination was the objective of this project. Two general approaches to achieving this goal were considered: (1) to analyze the garment itself and (2) to analyze swatches of the garments material that had been attached to the garment and are thought to he representative of the conditions to which the garment was exposed. While the first approach is best suited for incorporation into a field kit, the second approach was pursued during this feasibility study.

The desired sampling scheme was to be simple, non-obtrusive, and allow multiple swatches of the CPC material to be contaminated and decontaminated in a manner consistent with the actual garment. The swatches were to be placed on the garment prior to any site entry, decontaminated in-situ, and than removed and analyzed for the presence of chemical contamination. A successful swatch sampling scheme required that neither the edges nor the back of the swatch be exposed to the contamination or the decontamination procedure.

Two separate aspects of the swatch sampling scheme were identified: enclosure of the swatch so that the edges and back were not exposed, and attachment of the enclosure system to the garment. Several concepts for enclosure and attachment systems were conceived. The concept that appeared to best satisfy the criteria for an effective system used tape to integrate the requirements for enclosure and attachment. A tape-tab similar to a Band-Aid@ is fabricated from an adhesive tape and a swatch of the CPC garment material. The tab can be positioned anywhere on a garment.

A 1.5inch diameter swatch of material is made to face out of a 1.0-inch diameter exposure window that has been cut into a piece of tape. The tape-tab is completed by encasing the back of the swatch in another piece of tape. The adhesive on the tape creates a seal around the face, outside edge, and back of the swatch. The tab allows the swatch to be placed in intimate contact with the garment. The adhesive that remains outside the swatch serves to attach the tab to the garment.

The tape-tab system is lightweight, flexible, and appears to create a condition the best simulates actual garment performance. Swatches of CPC materials used to fabricate the tape-tabs can be requested from garment manufacturers or can be obtained by sacrificing garments. Garment manufacturers are receptive to supplying users with material swatches for the purposes of testing.

Several commercially available tapes were considered for this application (e.g., flame resistant, polyvinyl chloride (PVC), Teflon@, polyester, polypropylene, etc.). Teflon tape with a silicon adhesive was selected for the tape-tab based on its good chemical resistance, strength, flexibility, high/low temperature performance, and adhesion/release to a wide variety of CPC materials. Teflon/silicon tapes are available from Permacel and 3M. Permacel P-422 (2.0inch wide, Allied Resin) was selected for this study.

The performance of the tape enclosure system was judged by conducting chemical and physical adhesion tests. For the chemical tests, tabs were subjected to 15 minute full liquid immersions. Water, acetone, 30% sodium hydroxide, sulfuric acid, toluene, and methanol were selected as the test chemicals. After the exposure, the tab was patted dry and the swatch removed. The tab was considered effective if no chemical was visible outside the exposure window or on the back of the swatch.

Two types of physical adhesion tests were conducted. First, tabs were placed on a Level A garment and the garment was taken through a simulated use scenario that included flexing the tabs, kneeling and walking on the tabs, and rubbing the tabs against hazards such as walls. Under these conditions, the tabs stayed attached to the garment.

The second physical adhesion test focused on investigating the affect of the decontamination procedure on the integrity of the tape-tab. The decontamination procedure included spraying the tab with -60°C tap water, 50 strokes with a soft brush and a solution of water and liquid Tide®, and a final spray with tap water. The tape-tab was considered effective if no water was visible outside the exposure window after decontamination.

Several fabrication techniques were considered to satisfy the requirements for an effective tape-tab. The specifics of the final fabrication procedure are detailed in Appendix A.

Hardware --

The volatilization chamber had to be durable, easily opened and resealed, heat resistant, pressurizable, and allow for sampling. While sophisticated systems existed that satisfied these requirements, commercially available chambers are expensive and would still require some modification.

In lieu of modifying an existing system, a simple prototype volatilization chamber was fabricated according to Figure 1. Two, valved sampling ports were fitted into the lid of a 6-oz Mason jar. Also fitted into the lid was a thermocouple to monitor the inside temperature of the chamber and a holder for the material swatch. A Teflon stir bar is placed inside the chamber to facilitate mixing the air within the chamber.

The chamber is heated by suspending it in a water bath consisting of a 600 mL Pyrex® beaker and a hot plate/stirrer. The prototype chamber is inexpensive, simple, durable, easy to use, and effective (e.g., does not leak).

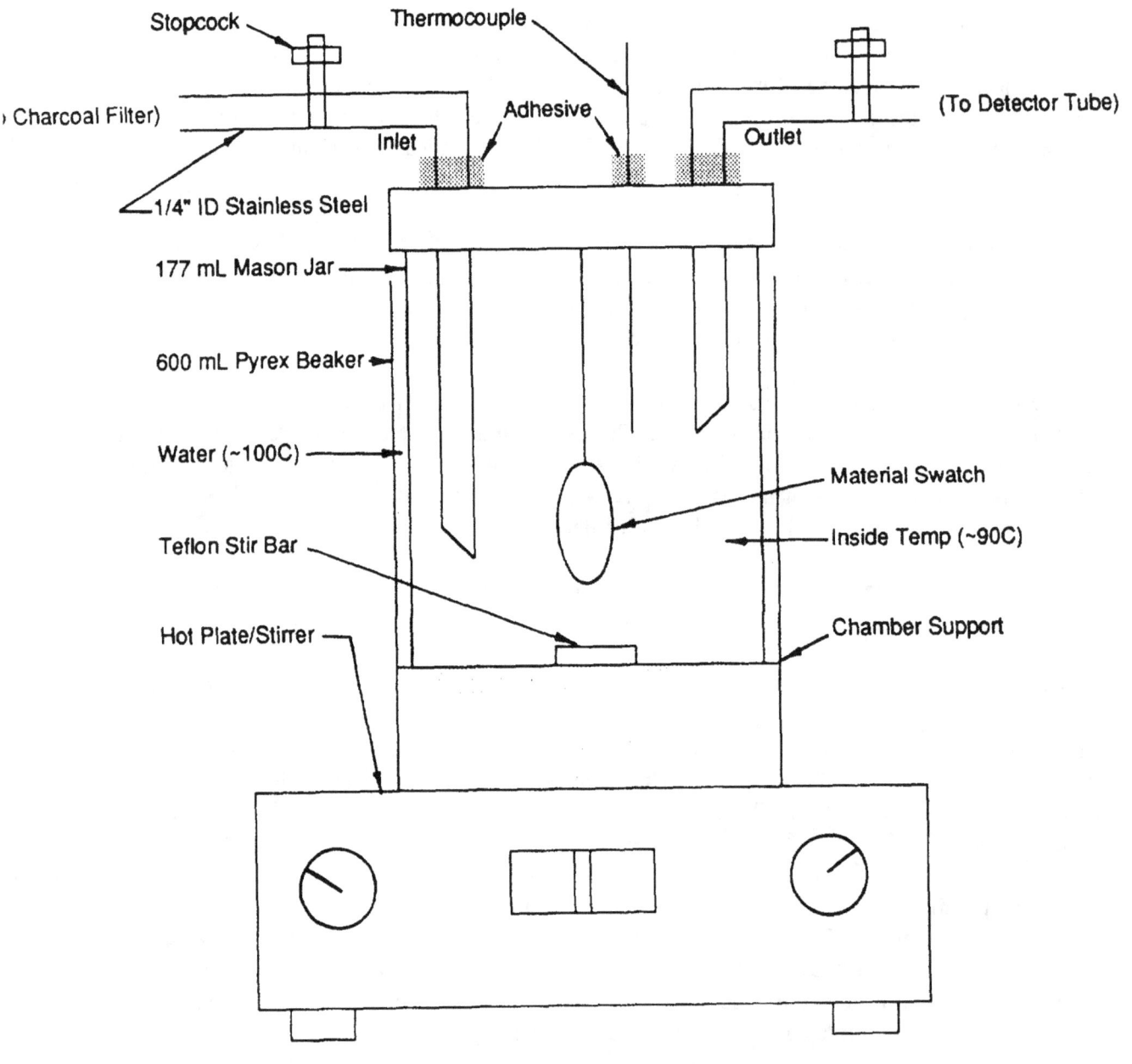

FIGURE 1. PROTOTYPE VOLATILIZATION CHAMBER.

DETECTOR TUBE TECHNIQUE

In its present form, the DT technique requires pre-positioning of protective clothing material swatches on a garment or destruction of the CRC. The procedural steps are summarized below. If a garment is being sacrificed for the test, the first two steps detailed below can be discarded In this case, a test swatch will be cut from the garment for use during Step 3.

- Fabricate a tape-tab from the material of interest and attach it to a garment,

- Expose and decontaminate the tab in-situ,

- Remove the material swatch from the tab and aerate for a specified time. (The aeration time can be specified by the tester, longer aeration times result in lower levels of chemical contamination. Theory suggests that all of the chemical adsorbed in a contaminated garment will desorb if sufficient time is allowed. The time necessary to achieve complete desorption is specific to the chemical/material combination and enhanced by increases in temperature.),

- Load the swatch into the volatilization chamber and heat to -90°C for a specified volatilization period. (The volatilization period like the aeration period must be matched to the chemical/material combination. One hour has been shown to be a sufficient volatilization period for many chemicals.),

A brief discussion of detector tubes and the sampling (bellows) pump is presented in Figure 2. The pump and detector tube form an integral unit; tubes and pumps from different manufacturers cannot be interchanged. The Draeger Gas Detection System was selected for this study based on the wide range of tubes offered by Draeger and their willingness to provide both technical and materials support to the project.

Briefly, DTs are narrow glass tubes packed with a series of reactive solid bed layers. The layer of interest is called the indicating layer and contains a chemical that causes a color change when reacted with the contaminate of interest. As can be seen in Figure 2C, the DT is calibrated in a series of indicating lines and in terms of concentration units. DTs are calibrated for either specific chemicals or classes of chemicals. In addition, each DT is calibrated for a specific sample volume. The required sample volume is indicated on each tube as "n", the required number of pump strokes (1 pump stroke = 100 mL).

A detector tube is used by snapping off the ends of the tube using the tip-breaker on the bellows pump, attaching the tube to the pump and the sampling port on the volatilization chamber, pulling the required volume of air through the tube, and than visually inspecting the DT for any color change. Care must be taken when orienting the detector tube in the pump. The proper direction of flow for each tube is marked on the tube with an arrow. The arrow should always face the bellows pump. DT responses are reported as the range of full indicating lines between which the length-of-stain ends for a specific test for this technique.

FIGURE 2. DESCRIPTION OF THE DRÄGER GAS DETECTOR

3 Description of the DRÄGER Gas Detector

The DRÄGER Gas Detector (Fig. 1) consists of the combination DRÄGER Tube + DRÄGER bellows pump. The DRÄGER Tube + DRÄGER pump must be used together.

Fig. 2A   DRÄGER Gas Detector, consisting of the gas detector pump and DRÄGER Tube

Brief description of the gas detector pump:

The gas detector pump is a hand-operated bellows pump (Fig. 2) This pump supplies 100 cm³ with each stroke. Thus, not only does the gas detector pump suck in the gas sample, but it also simultaneously carries out a volume measurement with each stroke. Its mode of operation is, therefore, that of a dosage pump

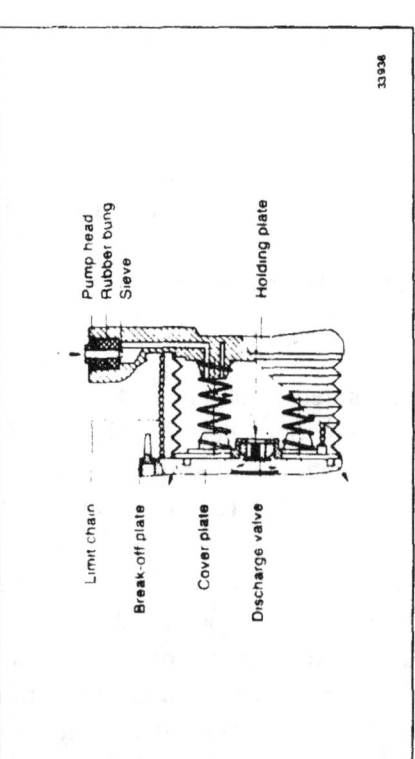

Fig. 2B Cross-section through the gas detector pump

The gas detector pump is made from neoprene and opens automatically after compressing and releasing the bellows. This opening process is effected by two steel springs built into the pump. The end of the suction process is reached when the limit chain is taut.

The gas detector pump has only one valve, which is closed when the gas sample is sucked in and opens again on squeezing the bellows. The pump head has an aperture into which the DRÄGER Tube to be used is inserted.

The time from releasing the bellows (after squeezing) until the limit chain is taut is termed the opening time of the gas detector pump: hence the opening time is the duration of one pump stroke.

The opening time depends on the flow resistance of the DRÄGER Tube inserted, which is a function of the filling preparation used. On the basis of the reaction kinetics, the flow resistance of the tube, and thus the opening time of the gas detector pump, differs depending upon the type of tube used. There are types of tube in which the pump opens in 3 seconds, but there are also tubes in which this process takes 40 seconds. However, the flow resistance for an individual type of tube varies only slightly, so that a range for the opening time of the gas detector pump can be kept to with each type of tube. These opening times are indicated in the instructions for use of the tubes concerned.

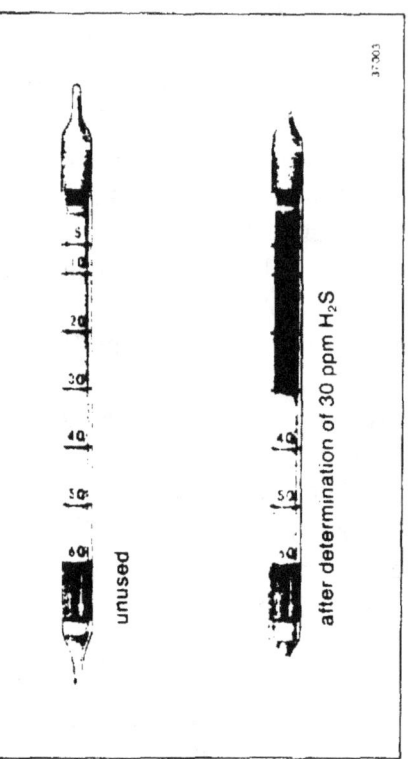

Fig. 2C   DRÄGER Hydrogen Sulphide 5/b Tube

(Source: Dräger Catalog, 1989)

DT responses are affected by temperature, humidity, pressure, chemical concentration, and chemical interferences. Furthermore, DTs have recommended storage conditions and shelf lives.

The concentration of chemical measured by a DT can be converted to a mass (ug) equivalent using the equation below. This equation was used during the comparative evaluation of the DT and analytical techniques.

$$ m = \frac{M \times A \times H}{10 \times 24} $$

where   m   =   mass of chemical detected, ug
       M   =   molecular weigh of chemical
       A   =   average concentration measured using DT, ppm
       H   =   number of pump strokes at which the detector tube is calibrated
       10,24 =   conversion factors

(Source: Detector tube measuring techniques, Leichnitz Ecomed-Verlagsqesellschaft Mbh, Federal Republic of Germany, 1983)

LABORATORY EVALUATION

Priority Chemicals --

The objective of the laboratory evaluation was to assess the feasibility of the DT technique to chemical/material combinations familiar to the fire service. The first task in designing the laboratory protocol was to develop a list summarizing the most often spilled chemicals. The following sources were referenced in developing the priority chemicals list:

- Interviews with the Sacramento, Prince George's County, Phoenix, and Del Ray Beach Fire Departments,

- ASTM Fl00l-89 - Standard Guide for Selection of Chemicals to Evaluate Clothing Materials,

- Synthetic Organic Chemicals. United States Production and Sales, 1988, USITC Publication 2219,

- Acute Hazardous Events Data Base, NTIS PB86-158946,

- "Development of a U.S. Coast Guard Chemical Response Suit", Report No. CG-D-16-87,

- "Material Development Study for a Hazardous Chemical Protective Clothing Outfit", Report No. CG-D-58-80, and

- The National Response Center.

The spill frequency data found in these references were similar. Omissions and inclusions of different chemicals from these lists are the result of differences in the definitions of "reportable" quantities of spilled chemicals. All of the data were used to compile the priority chemicals list shown in Table 1.

Table 1 contains 13 liquid and 2 gases considered during the laboratory test program. A selection criteria similar to that detailed in the U.S. Coast Guard Report No. CG-D-16-87 referenced above was used to select the priority chemicals. Briefly, the chemicals were selected based on their inclusion in one or more of the spill frequency lists, the requirement for use of and encapsulating suit (taken from the U.S. Coast Guard Report No. CG-58-80, referenced above), and their inclusion in ASTM F1001. This list represents a range of chemical types that are frequently encountered by fire department and other hazardous material response teams.

This list is not all encompassing or limiting to the applicability of the DT technique. The DT technique is applicable to any chemical for which detector tubes are available.

Priority Materials and Test Matrix --

The second task in designing the laboratory protocol was to select several protective materials commonly used in the fire service. Three CPC materials were selected: polyvinyl chloride (Standard Safety and Equipment), Viton®/Chlorobutyl (Life-Guard, Inc.), and Chemrel Max® (Chemron, Inc.). The first two materials have been in common use in the fire service for many years and represent a significant portion of its total encapsulating suit inventory. Chemrel Max is a relatively new barrier material that is gaining popularity in the fire service and other hazardous materials handling industries.

Table 2 summarizes the chemical resistance matrix for the priority chemicals and materials. The matrix was developed using manufacturers' data and the 3rd edition of the Guidelines For The Selection Of Chemical Protective Clothing. Table 3 summarizes a reduced chemical/material matrix selected for testing during the laboratory evaluation. This matrix was selected to provide a range of possible levels of contamination and to assess the feasibility of the method to several different classes of chemicals.

TABLE 1. FEMA PRIORITY TEST CHEMICALS

Chemical	Physical State	Encap. Suit Required	Spill Frequency	Detectable by Draeger	Detectable by Hach
Acetone *	liq	No	Yes	Yes	
Ammonia *	gas	Yes	Yes	Yes	
Chlorine *	gas	Yes	Yes	Yes	Yes
Dichloromethane *	liq	No	Yes	Yes	
Gasoline	liq	No	Yes	Yes	
Hexane *	liq	No	Yes	Yes	
Hydrochloric Acid *	liq	Yes	Yes	Yes	
Methanol	liq	No	Yes	Yes	
Nitrobenzene *	liq	Yes	Yes	(?)	
Sodium Hydroxide 1	liq	Yes	Yes	Yes	Yes
Sulfuric Acid	liq	Yes	Yes	Yes	
Tetrachloroethylene *	liq	No	Yes	Yes	
Tetrahydrofuran	liq	Yes	Yes	Yes	
Toluene	liq	No	Yes	Yes	
Vinyl Acetate	liq	No	Yes	Yes	

*   Indicates that chemical is also in ASTM F1001.

Requirement for encapsulating suit taken from USCG Report No. CG-D-58-80, "Material Development Study For A Hazardous Chemical Protective Clothing Outfit."

Spill Frequency taken from: 1) Interviews with Sacramento, Prince George's County, Phoenix, and DelRay Beach Fire Departments, 2) 1981-82 National Response Center Spill Frequency List, 3) NTIS Report No. PB86-158946, "Acute Hazardous Events Data Base," and 4) USCG Report No. CG-D-16-87, "Development of a U.S. Coast Guard Chemical Response Suit."

TABLE 2. CHEMICAL RESISTANCE MATRIX

Chemical	Breakthrough Time *		
	Viton/Chlorobutyl	Polyvinyl Chloride	Chemrel Max
Acetone	< 1 hr+	< 30 min	> 24 hr
Ammonia	> 2 hr	> 2 hr	NT
Chlorine	> 8 hr	< 1 hr	> 24 hr
Gasoline	< 20 min	< 15 min	NT
Hexane	> 3 hr	< 30 min	> 24 hr
Hydrochloric Acid	> 4 hr	> 3 hr	> 24 hr
Methanol	> 3 hr	< 1 hr	NT
Methylene Chloride	< 30 min	< 15 min	> 8 hr
Nitrobenzene	< 3 hr	> 2 hr	> 24 hr
Sodium Hydroxide	> 3 hr	> 6 hr	> 24 hr
Sulfuric Acid	> 3 hr	< 30 min	> 24 hr
Tetrachloroethylene	> 3 hr	< 15 min	> 8 hr
Tetrahydrofuran	< 20 min	< 15 min	NT
Toluene	> 3 hr	< 1 hr	> 8 hr
Ethyl Acetate	< 40 min	< 20 min	> 24 hr

* As reported by manufacturer or listed in 3rd Edition of Guidelines.
  Shaded boxes represent chemical/material combinations exhibiting breakthrough times < 1 hr.

*
+

14

	TABLE 3. TEST MATRIX		
Chemicals	Viton/ Chlorobutyl	Polyvinyl Chloride	Chemrel Max
Acetone	XX	XX	--
Dichloromethane	XX	XX	XX
Ethyl Acetate	X X	--	--
Gasoline	XX	--	--
Hexane	--	XX	--
Methanol	XX	XX	--
Nitrobenzene	--	XX	--
Toluene	--	--	XX

XX Tested.
--   Not tested.

Chemical Exposures --

The chemical/material combinations summarized in Table 3 were tested in triplicate. Chemical exposures were conducted by immersing a set of three tape-tabs, fabricated with the appropriate CPC material, in the test chemical for 15 minutes. After exposure, the tabs were removed from the chemical, blotted dry, and decontaminated according to the procedure detailed below under "Decontamination Procedure." After decontamination, the materials were aerated for a specified period and analyzed by either the DT technique detailed above under "DT PROCEDURE" or by the analytical technique discussed below under the heading, "Dynamic Thermal Stripping."

A three hour extraction time was used for those swatches analyzed by the DT technique. Triplicate tests were also conducted on swatches of unexposed virgin material to establish baseline data for both the DT and analytical techniques.

A limited study was also conducted to investigate the qualitative calibration/response of the DTs. A small aliquot of chemical was injected into a volatilization chamber using a syringe. The temperature inside the chamber was raised to -90°C and maintained for one hour. The volume inside the chamber was analyzed after the hour for the presence of the spiked chemical. Table B-l in Appendix B summarizes the qualitative calibration results.

Decontamination Procedure --

The decontamination procedure used during this study was designed to simulate a typical field decontamination scenario. While the overall time of this procedure may appear shorter than that typically used in the field, it was assumed that the times were representative of what an isolated area on a garment might encounter during decontamination.

- Rinse swatch for 30 seconds with room temperature tap water from a shower head,

- Wash swatch for 30 seconds using the decon agent. Washing consisted of scrubbing the swatch with a soft bristle brush. The decon agent was a 12 gm/L, solution of liquid Tide@ and water,

- Rinse swatch for 30 seconds with room temperature tap water,

- Pat swatch dry with a paper towel to remove all visual signs of wetness,

- Remove swatch from tab enclosure and aerate for 21-24 hours by hanging swatch in fume hood.

Dynamic Thermal Stripping --

The results from the DT technique were compared to data generated using an analytical technique called Dynamic Thermal Stripping (DTS). DTS is a technique wherein chemical contaminates present in a specimen are desorbed into a stream of hot gas. The now contaminated gas stream passes through an activated carbon filter which concentrates the chemical. Subsequently, the contaminating chemical is desorbed from the filter and analyzed.

Hot helium (100°C) was blown past the material swatches at 6 mL/min for 60 minutes during this study. An Envirochem Model 1260 dynamic thermal stripper was used with Tenex® activated carbon adsorption tubes.

After 60 minutes, the adsorption tubes were loaded into an Envirochem Unicon Series 810 Thermal Desorber and desorbed into a stream of 230°C helium flowing at 1.6 mL/min. The helium stream was fed into an HP Model 5890A gas chromatograph (GC) fitted with a 30-meter, DB1 capillary column (0.25um film thickness, Jane W. Scientific), Final detection was via a flame ionization detector (FBI) in the GC. The output from DTS was the weight (ug) of chemical contaminate desorbed from the material swatch.

Calibration and test data for DTS are presented in Appendix C. Due to confusing data gathered early on in the investigation (See discussion under "Results" below), the data presented in Appendix C are only for those chemical/material combinations used in the comparative evaluation discussed below. Three point calibration curves were used for the DTS technique. The calibration standards were made via series dilutions. Aliquots of the calibration standards were injected directly into the GC and the associated responses (areas) recorded.

Only a small piece of each swatch (approx. 1/20th) was analyzed by DTS so as not to overload the GC column. The amount of chemical measured in the small samples was scaled-up to determine the amount of chemical present in the entire swatch.

Effect of Decontamination on Chemical Resistance --

A limited study was conducted to investigate the effect of decontamination on the chemical resistances of the test materials. Permeation cup tests were performed on swatches of new and decontaminated specimens to evaluate this affect. Testing was conducted in accordance with the February 1990 draft of the ASTM permeation cup test method. Weight losses from the cups were measured at 60 minutes. Each chemical/material/condition was tested in triplicate. The chemical/material combinations investigated were ethyl acetate/Viton/Chlorobutyl, acetone/PVC, and toluene/Chemrel Max.

FIELD EVALUATION

The purpose of the field study was to obtain information on the potential effectiveness, usefulness, and acceptability of the DT technique in field situations. The study was conducted on October 18-19, 1990 at Arthur D. Little's Cambridge facility. The two-day study included two members of the Cambridge, Massachusetts Fire Department Hazardous Material Response Team and one member of the Phoenix Fire Department Special Operations Branch.

The study included contamination, under simulated field conditions, of tape-tabs of CPC materials mounted at several locations on splash suits, decontamination of the suits, and subsequent analysis of the tabs for the presence of chemical contamination using the DT technique. Ethyl acetate was selected for the study based on its ease of handling, inclusion in ASTM F1001, and prior experience with the chemical during the laboratory evaluation. Five protective materials were selected for the field study, three CPC materials and two materials used in the fabrication of firefighter turnout gear (TOG). Neoprene, PVC, and Viton/Chlorobutyl were the CPC materials. Leather and Nomex III® were the turnout materials.

Five to six tape-tabs were fabricated from each of the CPC test material and applied to a manikin donned in a splash suit. The manikin was sprayed with ethyl acetate in a random manner for 5 minutes using a garden type sprayer. This exposure scenario attempted to simulate an uncontrolled release such as a ruptured pressurized vessel.

The manikin was exposed three times during the study. Only one type of CPC material was attached to the manikin during each exposure. One swatch of the leather material was tested during the PVC exposure and one swatch of the Nomex III during the Viton/Chlorobutyl exposure.

After contamination, the manikin was moved to a decontamination pool, decontaminated according to a field decon procedure similar to that used during the laboratory evaluation, and then set aside. After decon, the swatches were removed from the tabs and aerated for one hour. Aeration was conducted outside, average temperatures were 75°F during the neoprene exposure and 60°F during the PVC and Viton/Chlorobutyl exposures. While not measured during this study, windspeed was high both days.

After aeration, the swatches were loaded into volatilization chambers and maintained at -90°C for one hour. After the extraction period, the volumes inside the chambers were analyzed using DTs. Analyses were also conducted, in triplicate, on virgin swatches of the CPC materials.

The conditions expected in the field were also reproduced in the laboratory in an attempt to generate comparative data. In the laboratory, the tape-tabs were exposed to ethyl acetate using the garden sprayer for 7.5 minutes. The decontamination procedure used was the same as that used during the laboratory evaluation. Each material was tested in duplicate.

The field study was documented photographically and on video tape. The protocol for the field study is included at the end this report as Appendix D.

RESULTS AND DISCUSSION

PRELIMINARY RESULTS AND GENERAL OBSERVATIONS

DTs are designed to sample a small portion of a large air volume exhibiting a relatively constant concentration. In the technique developed in this study, however, the DTs are used to analyze an air sample with a concentration that decreases with each additional pump stroke. Since DTs are highly engineered for their designed purpose, we conducted hundreds of tests to assess their suitability to the application of interest here.

Preliminary investigations focused on the effect of volatilization chamber size and the number of pump strokes on DT responses. Hexane was selected for these tests. The results of the preliminary tests are presented in Appendix B, Tables B-2 and B-3. These tests were successful in demonstrating that DTs could be used to detect the presence of a range of chemical contaminates in several CPC materials, however interpretation of the results proved difficult. DT responses appeared to increase with increasing pump strokes and decreases in the size of the volatilization chamber. Furthermore, the DT and DTS results did not compare consistently (i.e., DTS did not always produce more sensitive results than DT as expected).

The effects noted above can be related in part to the efficacy of quantitatively evacuating the DT volatilization chamber. For a large chamber and a low number of pump strokes, all of the chemical present in the volatilization chamber may not be drawn through the DT. This will result in an unrealistically low DT response. If on the other hand a large number of pump strokes is used with a small chamber, all of the chemical present in the chamber may be drawn through the DT early on with the final pump strokes drawing only clean air through the tube. Drawing uncontaminated air through a tube that already exhibits a response can cause the length-of-stain to either disappear or migrate unrealistically to a higher response.

The effects discussed above complicate comparing results quantitatively since each DT is calibrated for a different number of pump strokes. To avoid confusion, the decision was made to minimize the size of the volatilization chamber, use DTs at their calibrated number of pump strokes, and compare results on a relative basis only.

The results of the qualitative calibration study (see Table B-l, Appendix B) support the conclusion that this technique should only be considered on a relative basis. The actual results for the seven chemicals tested were all below that expected based on the amount of chemical injected into the chambers. This result was not unexpected, DT accuracies are typical reported as + 25-100% by both manufacturers and users.

LABORATORY EVALUATION

The primary objectives of the laboratory evaluation were to assess the feasibility of the DT technique to chemical/material combinations familiar to the fire service, and to compare the DT results with those generated using an established laboratory technique. Both objectives were met during this study.

The preliminary laboratory data summarized in Table B-1 in Appendix B verifies the conclusion that CPC materials that are exposed to a chemical become contaminated and that the level of contamination is specific to the chemical/material combination. MeCl will be used as an example. The DT technique was used to demonstrate that the residual level of chemical contamination after equal exposure/decontamination scenarios was highest in PVC, intermediate in Viton/Chlorobutyl, and lowest in Chemrel Max, on a relative basis. This conclusion is consistent with the relative chemical resistances of these materials as shown in Table 2.

The preliminary tests demonstrated the applicability of the DT tube technique to three CPC materials and the following seven classes of chemicals often faced by the fire service: ketones, halogenated hydrocarbons, esters, normal alkanes, alcohols, aromatics, and mixed hydrocarbons.

The second objective of the laboratory evaluation was to compare the DT results with those generated using an analytic method. As detailed under "Methodology", CPC swatches were prepared, cut into two equal halves, and than analyzed by either DT and DTS. Table 4 summarizes the results of the comparative evaluation. Similar or overlapping results were obtained for all but one of the four chemical/material combinations tested. A greater amount of acetone was measured in the PVC by DTS than by DT. This result was not unexpected since DTS exhibits improved sensitivity to acetone than does the DT technique.

The closeness of the DT and DTS results and the demonstrated applicability of the procedure to a range of chemical classes and CPC materials lead to the conclusion that the DT technique is a useful and effective method for assessing the presence of chemical contamination in CPC materials.

The final portion of the laboratory evaluation focused on determining the effect of the decontamination procedure on the chemical resistance of CPC materials. Triplicate permeation cup tests were conducted on virgin and decontaminated swatches of material according to the most recent draft of the ASTM method. Table 5 summarizes the results of this limited investigation. No significant differences in cumulative weight losses at 60 minutes were measured for the three chemical/material combinations tested. This result was not unexpected for the decontamination procedure used during this study. Caution is recommended in interpreting these results since previous studies have shown that the effect of decontamination on chemical resistance can be significant depending on the decontamination technique selected.

TABLE 4. COMPARATIVE TEST RESULTS				
Test	Chemical	Material	Analysis	ug Detected
1	Acetone	PVC	DT	4840- 14520
				14520-29040
				4840- 14520
			average =	6615-19360
2			DTS	19839
				36082
				31932
			average =	29284
3	Methanol	Viton/Chlorobutyl	DT	<66.4
				<66.4
				<66.4
			average =	<66.4
4			DTS	672
				47
				28
			average =	38
5	Methylene chloride	Viton/Chlorobutyl	DT	<708
				708-2124
				708-2124
			average =	708-1652
6			DTS	1253
				1437
				1821
			average -	1504
7	Gasoline	Viton/Chlorobutyl	DT	780-2340
				2340-3900
				780-2340
			average =	1300-2860
8			DTS	1485
				1388
				1605
			average =	1493

TABLE 5. PERMEATION CUP TEST SUMMARY			
		Cumulative Weight Loss (gm)	
Chemical	Material	Virgin Material	Deconed Material
Ethyl Acetate	Viton/Chlorobutyl	0.029±0.007	0.030±0.0003
Acetone	PVC	2.168±0.131	2.165±0.124
Toluene	Chemrel Max	0.002±0.002	0.001±0.001

FIELD EVALUATION

The feasibility of the DT technique to field situations was successfully demonstrated during the field evaluation. The participants in the field study found the procedure easy to use, and the results easily interpretable and of immediate benefit. Both the participants and Steve Storment strongly supported application of the method to materials other than CPC, specifically, firefighting turnout gear. The major comments made during the study are detailed below. Mr. Steve Storment prepared a short summary of the field study which is presented at the end of this report as Appendix E.

- The DT technique could be used as a cost-effective means of saving suits (both CPC and TOG) that would otherwise be considered contaminated,

- This technique could be used to reduce disposal costs by aiding in decisions regarding whether or not a contaminated suit can be disposed of as hazardous or nonhazardous waste,

- This technique could be easily and successfully conducted by any company and by any individual within a company; however, the participants commented that designating the task to only a few individuals within a company would ensure consistent results.

A short video was prepared on the study that includes a discussion of the project, the objectives of the field study, an overview of the DT analysis technique, and actual field footage. The video ends with comments made by the participants and Mr. Steve Storment. The data generated during the field study are discussed below.

The results for the baseline tests conduced in the laboratory prior to the field study are summarized under Lab Results in Table 6. Average tube responses for all of the virgin specimens were below the minimum detectable limit of the ethyl acetate detector tube. Average responses for neoprene, PVC, and Viton/Chlorobutyl were >3000 ppm, >3000 ppm, and 400-1000 ppm, respectively.

TABLE 6. FIELD RESULTS - ETHYL ACETATE

Material	Lab Result?	Detector Tube Response (ppm)* Tab Position					
		1	2	3	4	5	6
Neoprene	>3000	<200	400-1000	400-1000	400-1000	200-400	1000-2000
Polyvinyl Chloride	>3000	<200	400-1000	<200+	>3000	>3000	1000-2000
Viton/Chlorobutyl	400-1000	**	200-400	<200++	<200	<200	200-400

†   Lab results generated under simulated field conditions: 7.5 minute chemical spray at -80°F. Field results generated using a 5 minute chemical spray. Field temperatures were -75°F for neoprene and -60°F for PVC and Viton/Chlorobutyl.

*   Background analyses were conducted in duplicate on virgin swatches of each test material. No response was recorded for either of these tests.

+   Swatch taken from leather knee pad of firefighter turnout gear.

**   Extraction chamber was opened prematurely invalidating test result.

++   Swatch of Nomex III taken from firefighter turnout gear.

23

One to six swatches of each material were exposed/deconed/analyzed. The detector tube analysis technique differentiated the performances of different materials and differentiated the residual level of chemical contamination on different parts a single garment.

The columns labeled "Tab Position" in Table 6 refer to the positions of the tabs on the suit as shown in Figure 2 in Appendix D. Using PVC as an example, the lowest level of contamination was measured in tab 1 which was positioned on the back of the manikin and the highest level of contamination was measured in tabs 4 and 5 which were positioned at the ankle and hip of the manikin. These results are consistent with expectations. Only the front of the manikin was exposed during the test, therefore, tab 1 was never directly exposed and one would not expect to measure any chemical in this swatch. Tabs 4 and 5 were located at the two lowest positions on the manikin and one would expect these tabs to see the highest level of contamination as a result of a combination of direct contact during spraying and secondary contact from the run-off of chemical sprayed above these tabs.

Overall, the field and laboratory results compare well. A response greater than 3000 ppm of ethyl acetate was detected in the laboratory for PVC swatches exposed to a direct spray for 7.5 minutes at 75°F. The lower concentrations measured in the field are the result of decreased ambient temperature, and a decrease in the duration of direct spray. The spray pattern used in the laboratory concentrated on the swatches; a random spray pattern was used in the field thus decreasing the duration of direct spray contact to any one area on the garment. The potential for chemical to absorb into a swatch is proportional to the duration of chemical contact (e.g., duration of direct spray). The results for neoprene and Viton/Chlorobutyl are similar to those for PVC.

The DT analysis technique has application not only to CPC but also to turnout gear. Swatches of leather and Nomex III were tested during the field study to demonstrate this application. The results for these materials are also included in Table 6. No residual contamination was measured in either the leather or the Nomex III. Similar tests were conducted in the laboratory which support these results. This information was of great interest to the participants. They commented that firefighters wear TOG in excess of 90% of the time while wearing CPC less than 10% of the time. Therefore, any test method developed to assess the presence of chemical contamination in protective clothing must necessarily be applicable to TOG.

APPLICATION OF DT TECHNIQUE

The applicability of the DT technique has been demonstrated in the laboratory and the field. While not optimized for field use, the technique is both useful and effective, and could be used in its present form to assess the level and location of chemical contamination in protective clothing. All necessary equipment can be either purchased off-the-shelf or can be easily fabricated in-house. The approximate cost for a DT system containing three volatilization chamber assemblies is $825, not including expendables (see Table 7). The approximate cost for expendables required to conduct testing of a single garment (we recommend ten swatches per garment) is $75.

24

TABLE 7. DT SYSTEM PARTS COSTS		
Item	Approximate Cost	Source
TAPE-TAB		
Specimen Stencils (1.0"& 1.5" ID)	$ 70 two required, $35 each	1
Permacel P-422, Telflon Tape	$ 25 per garment	2
Material Swatches	No charge	Clothing Mfg.
VOLATILIZATION CHAMBER		
Mason Jar (6 oz.)	$ 3 @ $12 per dozen	Grocery Store
Stir Bar, small round	$ 9 @ $3 each	3
Heavy duty glass beaker, 600 mL	$ 21 @ $7 each	3
Hot plate/stirrer	$ 4 5 0 @ $150 each	3
Thermocouples and meter	$ 50	3
Stainless Steel Tubing, 1/4" ID	$ 3 @$1 perfoot	4
Tygon Tubing, 1/4" ID	$ 3 @ $1 perfoot	3
Teflon Tubing, -1/4" ID	$ 6 @ $2 per foot	4
Charcoal Filter	$ 9 per chamber/garment	3
Stopcocks	$ 84 @ $14 each	5
DT ANALYSIS		
Sample Pump, Drager Bellows Pump w/counter	$ 125	6
Detector Tubes	$30-50 per garment	6

Approximate cost for a 3-chamber system = $825 without expendables.

Approximate cost for expendables (highlighted in grey) consumed during analysis of one suit (10 tabs) = $75.

sources:

1. The Hudson Die Group, (508) 559-7370
2. Allied Resin, (617) 337-6070
3. Fisher Scientific, (412) 562-8300
4. Cole-Parmer Ind. Co., (800) 323-4340
5. VWR Scientific, (415) 467-4100
6. BGI Incorporated, (617) 891-9380

PROTECTIVE CLOTHING REUSE CONSIDERATIONS

The principal purpose of this study was the development of a procedure for assessing the extent of chemical contamination of protective clothing. Such a procedure would become part of an overall process for using and disposing of the clothing. This process is illustrated in Figure 3, and described below. Many of the decision points in the process are influenced strongly by financial considerations. If money were no object, then it is likely that the decision would always be to use clothing once and discard it. There would be no incentive to consider questions as:

- What is the real cost of reusing protective clothing?

- How many times must an expensive garment be reused before its cost per use becomes less than the cost of a single-use garment?

A preliminary approach to answering such questions is also addressed below.

Clothing Use/Disposal Process --

All clothing items, whether they are classified as single-use or reusable, will be used and eventually discarded. One view of the process is shown in Figure 3 and encompasses the following steps:

Step 1 - Selection. Prior to using the clothing it must be properly selected.

Step 2 - Use.

Step 3 - Decontamination. The three major reasons that clothing is decontaminated are: (1) to minimize the transfer of contamination from the work site; (2) to reduce the contamination to a level that would allow the clothing to be disposed of as a non-hazardous waste; and (3) to reduce the contamination to a level that would allow safe reuse of the clothing. The decontamination costs can differ significantly depending on which of the three reasons is the desired goal. There are a variety of decontamination methods; most involve scrubbing the surface of the garment with water/detergent solution, rinsing, and air drying. The U.S. Environmental Protection Agency describes a 19-step decontamination process in its handbook for Hazardous Materials Incident Response Operations.

Step 4 - Assess Decontamination. The effectiveness of the decontamination process is measured as one input into the decision of Step 5.

Step 5 - Reuse/Discard Decision. If the clothing was purchased as a single-use item, go to Step 8. If there is the desire to reuse the clothing, then this decision will consider the success of the decontamination process as well as the condition of the garment (Step 6).

26

FIGURE 3. CPC REUSE/DISPOSAL DECISION LOGIC

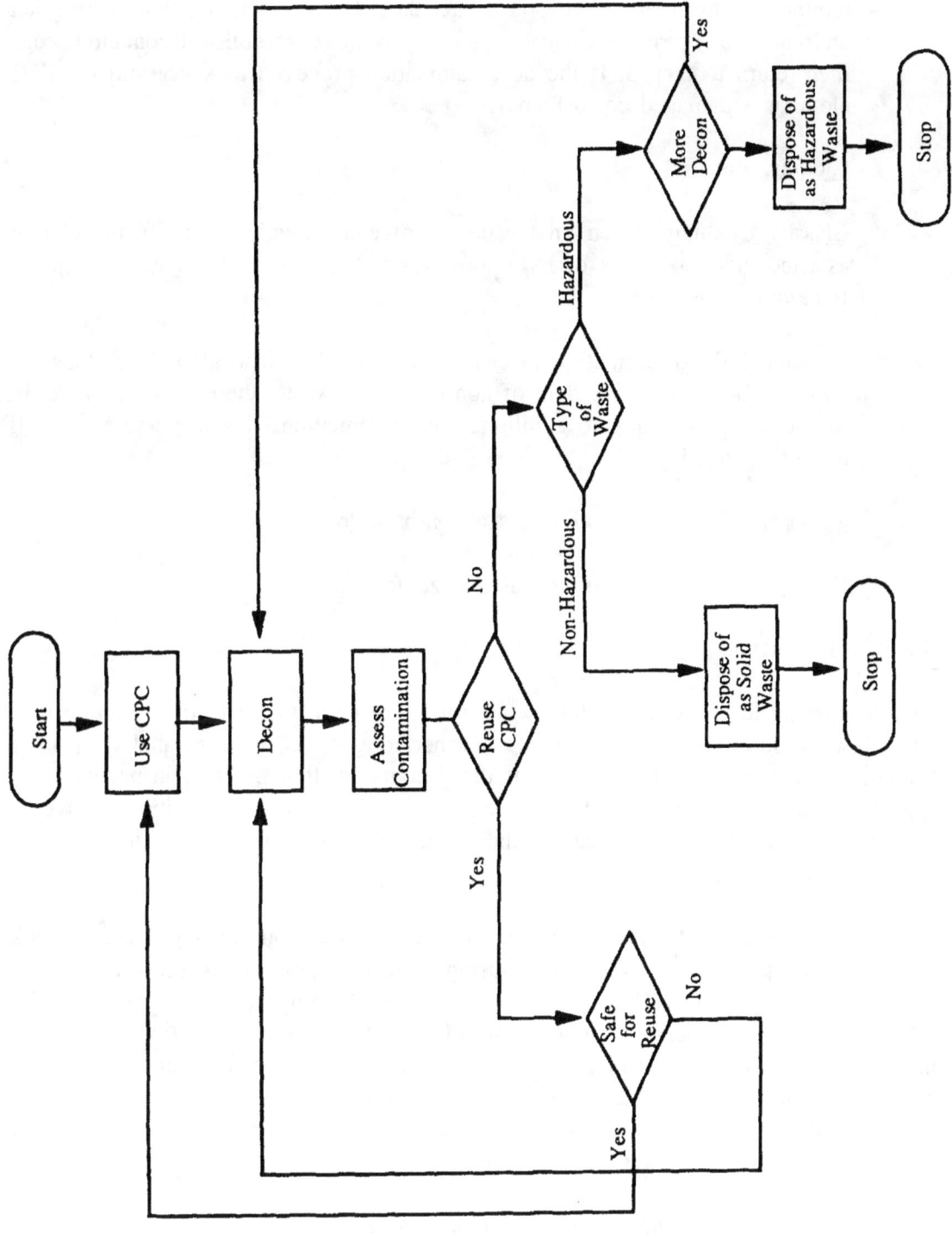

Step 6 -   Clothing Inspection. Is the clothing abraded, torn, punctured, cracked, etc. and how much would it cost to thoroughly inspect, repair, and maintain the garment? If the decontamination process has been successful and the clothing retains its physical and chemical barrier integrity, then the clothing would be returned to. inventory for use at a later date. If the clothing retains its physical and chemical barrier integrity but has not been successfully decontaminated, then return to Step 3. If the decontamination process was successful but the clothing is damaged go to Step 7.

Step 7 -   Clothing Repair.

Step 8 -   Classify Clothing. Based on the degree of contamination, classify the clothing as either non-hazardous or hazardous waste. If non-hazardous, go to Step 10. If hazardous, go to Step 9.

Step 9 -   Additional Decontamination Decision. Because the disposal of hazardous waste is more expensive than of non-hazardous waste, there may be the desire to subject the clothing to additional decontamination. If yes, go to Step 3. If no, go to Step 11.

Step 10 -   Disposal. Dispose of clothing as a non-hazardous waste.

Step 11 -   Disposal. Dispose of clothing as a hazardous waste.

Financial Considerations --

Readily evident from the above discussion is that total cost of using protective clothing is greater than just the purchase cost of the item. Furthermore, the costs associated with decontaminating, assessing the effectiveness of the decontamination process, inspecting, repairing, storing, and discarding the clothing can be significant. These activities involve labor and materials that must be considered when comparing the total costs of single- versus multiple-use protective clothing.

As part of a recent assignment for the U.S. Environmental Protection Agency, we undertook a preliminary effort to quantify the total cost of using protective clothing. In the study, "Preliminary Assessment of Life-Cycle Costs of Protective Clothing" (EPA Report No. PB90-219171/AS), estimates were derived for the costs of decontamination, inspection, maintenance, storage, and disposal of protective clothing. The cost of decontamination was estimated based on the EPA's 19-step process referred to above. This process is intended as a surface decontamination process and may or may not be effective for the removal of matrix contamination.

The estimated costs were combined with assumptions on the number of times the clothing would be used and the number of items of clothing used per response to yield a mathematical model that estimates the cost/use for the clothing. The cost/use value enables comparison of the economics of various usage scenarios. For example, when is it less expensive to use multiple, single-use garments than to reuse a single, more expensive garment? Because of certain fixed costs of decontamination and reuse, is there a garment purchase cost below which the garments should be considered single-use?

In the present study, the mathematical relationship that was developed for the EPA was expanded to include the cost of assessing the presence of contamination using the technique &scribed above. Estimates for the cost of the materials used in assessing decontamination were taken from Table 7; we also estimated the labor hours that might be required to perform the technique. The estimates and the underlying assumptions are as follows:

- Non-expendable equipment for one complete set-up is estimated at $825. On average, we estimate that normal wear would require replacement of this equipment after it had been used for 100 garments. Thus, the cost for non-expendable equipment would be $8.25 per garment.

- We assume that 10 patches (tabs) would be placed on each garment. Thus, ten clothing specimens would be tested per garment. The cost for the tape for the ten tabs is estimated at $25. The cost for the ten length-of-stain tubes for the analyses is estimated at $40. The cost for three carbon filters is estimated at $9. Thus, in total, there would be $75 of expendables per garment.

- We estimate one hour will be required to set-up and take-down of the equipment.

- We estimate that six hours of labor will be required to perform the ten analyses for one garment.

The EPA model was also modified to include costs for decontamination, assessment of decontamination, and inspection after each use rather than after each use except the last one. Using these assumptions, we estimated the cost/use for:

- four garment purchase prices: $300, $750, $1500, and $3000.

- one daily garment usage rate: 7 garments per day.

- two hourly labor rates: $0 and $25. The $0 value assumes that the persons are already being paid for their time and that the time is available for the activities described above.

- three number of uses: 2, 5, and 10; i.e., the clothing would be used twice, five times, or ten times.

- two starting conditions: one that the garments have already been paid for (i.e., the present stock of the fire service); the other that new garments must be purchased.

29

The results are summarized in Table 8. Each entry in the table is an estimate of the cost/use of a garment for the indicated combination of initial purchase cost, and the number of times any single garment is used. For example, $100 is the estimated cost per use for a garment with a purchase cost of $300, that is used twice when the garments have already been paid for, and for which there is no cost for labor. For the same conditions except a labor cost of $25/hr, $275 is the estimated cost per use. Observations on the table include:

• Labor at $25/hr adds about $200/use to the total cost of a garment, compared to labor at $0/hr.

• For garments already in inventory,

  – If there is no charge for labor, the cost/use is $100-150 for essentially all use scenarios.

  – At a labor rate of $25/hr, the cost/use ranges from $275-350 for essentially all use scenarios. The approximate $200 increase over that for the $0/hr scenarios is due largely to the labor cost of assessing garment contamination. Rather than incur this cost, a better approach might be to assume the garment is contaminated, discard it, and purchase a new garment. Thus, the uncertainties of decontamination are avoided an the firefighter gets a new garment for essentially the same cost.

  – As the purchase price of the garment increases to $3,000, the more economical solution is to decontaminate and reuse the garment.

• When considering the purchase of new garments,

  – Three or more uses are required to reduce the cost/use of a $3,000 garment to below $1,500. Similarly, three or more uses are required to reduce the cost/use of a $1,500 garment to below $750. These results exemplify the case in which it may be more cost-effective to use multiple, new, less-expensive garments than to reuse the more expensive garment multiple times. As the purchase costs of the garment decreases, this trend becomes more pronounced since the fixed costs associated with labor, decontamination, etc. overwhelm the purchase cost of the garment.

  – The cost/use drops significantly as the number of uses for a single garment increases to ten. As the number of uses required to "break even" increases, however, questions arise as to the capability of a garment to maintain its integrity over a large number of uses.

  – For any set of conditions under which the cost/use of multiple uses of a garment is greater than the initial purchase cost of the garment, one might consider the garment as single-use.

30

TABLE 8. ESTIMATED COST/USE FOR PROTECTIVE CLOTHING* ($, rounded to nearest $25 or $50)					
		Purchase Cost Not Included		Purchase Price Included	
Purchase Price $	Total Number of Times Garment Is Used	Labor Rate $0/hr	Labor Rate $25/hr	Labor Rate $0/hr	Labor Rate $25/hr
300	2	100	275	250	425
	5	100	275	150	325
	10	100	275	125	300
750	2	100	300	500	675
	5	100	300	250	450
	10	100	300	175	375
1500	2	125	300	875	1050
	5	100	300	400	600
	10	100	300	250	450
3000	2	150	350	1650	1850
	5	125	325	725	925
	10	100	325	400	625

*   Based on the use of 5 to 10 garments per day.

31

REFERENCES

Anderson, Dr. C.P., memo to American Society for Testing and Materials F-23.30.04 Task Group, "R & D Center Participation in Decontamination Round Robin," June 1987.

Bentz, A.P., memo to American Society for Testing and Materials F-23.30.04 Task Group, "Round Robin Test Method for Measuring Matrix Contamination (Decontamination Study)," 19 June 1987.

Ehntholt, D.J., and D.L. Cerundolo, "Evaluation of Decontamination Agents and Methods for Removing Contaminants from Protective Clothing," Phase I Draft Report, U.S. Environmental Protection Agency, Contract No. 68-03-3293, September 1986.

Garland, C.E., "Chemical Contamination and Decontamination of Protective Clothing," presented originally at the 1985 Conference of the Association of College Professors of Textiles and Clothing - Western Division, Naper, CA, October 24, 1985. This version is updated to include more recent experimental data.

Kairys, C.J., and 2. Mansdorf, "Decontaminating Protective Clothing Offers Challenge to Manufacturers," Occupational Health & Safety, (59)5:36-42, May 1990.

Kominsky, J.R., and E.T. McIlvaine, "Decontamination of Fire Fighters' Protective Clothing with Trichlorotrifluoroethane," paper presented at the American Industrial Hygiene Conference, Las Vegas, Nevada, May 19-24, 1985.

Moore, J.A., "Survey of Use and Maintenance Procedures for Chemical Protective Total Encapsulation Garments," Performance of Protective Clothing, ASTM STP 900, R.L. Barker and G.C. Coletta, Eds., American Society for Testing and Materials, Philadelphia, PA, 1986, pp. 286-297.

"Proper Maintenance Will Reduce the Effects of Contamination to Fire Fighters," Winsol Laboratories, Inc., 1417 NW 51st Street, Seattle, Washington, 98107 (Revised July 1988).

Sarner, S.F., and N.W. Henry, III, "The Use of Detector Tubes Following ASTM Method F-739-85 for Measuring Permeation Resistance of Clothing," Am. Ind. Hyg. Assoc. J., 50(6): 298-302 (1989).

Schwope, A.D., and D.E. Ehntholt, "Chemical Protective Clothing Contamination, Decontamination, and Reuse," prepared for the Federal Work Group on Protective Clothing, U.S. Environmental Protection Agency, Contract No. 68-03-3293, December 1988.

Smith, I.D., and K.E. Burke, "Decontamination of Protective Suit Materials," Performance of Protective Clothing: Third Symposium, ASTM STP 1037, J.L. Perkins and J.O. Stull, Eds., American Society for Testing and Materials, Philadelphia, PA, 1989.

Vahdat, N., and R. Delaney, "Decontamination of Chemical Protection Clothing," Am. Ind. Hyg. Assoc. J., 50(3): 152-156 (1989).

APPENDIX A

TAPE-TAB FABRICATION TECHNIQUE

1. Die cut a 1.5-inch diameter swatch from the protective clothing material of choice.

2. Cut a -5-inch long strip of tape and lay it adhesive side up. Fold over -1/2-inch of one end of tape to form a tab. The tab will be used to remove the tape-tab from the release sheet and garments.

3. Die cut a 1.0-inch diameter exposure window in the center of the tape.

4. Center the material swatch over the exposure window and apply to tape. Ensure that the normally outside surface of the material is made to face out of the exposure window.

5. Cut a -3-inch long strip of tape and apply it (adhesive to adhesive) crosswise and centered over the swatch. Approximately 1/2-inch of the tape will extend over the edges; fold these over onto the front of the tab ensuring that they do not overlap any part of the exposure window.

    Rub the back of the tab to ensure good adhesion between the tapes and the swatch.

6. Apply the tape-tab to a Mylar@ release sheet.

7. Rub the tape that covers the swatch to ensure good adhesion.

8. Tape-tabs can be fabricated at any time prior to use. It is not recommended that tabs be fabricated at a response. Pre-fabricated tape-tabs (on Mylar sheets) can be placed in Zip-Lock@ bags and stored in the freezer. Colt storage will maintain the adhesive properties of the tape-tab.

APPENDIX B

QUALITATIVE CALIBRATION AND PRELIMINARY DT TEST RESULTS

TABLE B-1. QUALITATIVE CALIBRATION RESULTS

Chemical	Spike (ul)	Pump Strokes	Response Expected	Response Actual
Acetone	25	10	8140 (6000-9000)[+]	1000-3000 3000-6000 3000-6000
[Indication range: 100-1000-3000-6000-9000-12000]				
Ethyl Acetate	12	20	1471 (1000-2000)	400-1000 400-1000 2000-3000
[Indication range: 200-400-1000-2000-3000]				
Gasoline (unleaded)	1	2	?[†]	<100 <100 <100
[Indication range: 100-300-500-1000-1500-2000-2500]				
Hexane	8	6	2450 (2000-2500)	300-500 700-1000 300-500
[Indication range: 100-300-500-700-1000-1500-2000-2500-3000]				
Methanol	1	5	1186 (1000-2000)	500-1000 1000-2000 500-1000
[Indication range: 50-100-300-500-1000-2000-3000]				
Methylene Chloride	2	10	750 (500- 1000)	100-300 300-500 300-500
[indication range: 100-300-500-1000-1500-2000]				
Toluene	0.5	5	225 (200-300)	<50 <50 <50
[Indication range: 50-100-200-300-400]				

[+] Expected response is presented as: calculated response (full indicating range that calculated response falls within).

[†] This tube is calibrated for n-octane; the concentration of n-octane in the gasoline was not known.

TABLE B-2. EFFECT OF CHAMBER SIZE
(8 ul Hexane Spike)

Chamber Size (ml)	RESPONSE (PPM) No. of Pumps				
	2	4	6	8	10
473	100-300 100-300	500 300-500	500-700 500-700	700-1000 700-1000	700-1000 900
177	-- -- 500-700 --	-- -- 1000-1500 --	-- -- 1000-1500 --	-- -- 1000-1500 --	1500 1000-1500 1500 1500

TABLE B-3. EFFECT OF PUMP STROKES
(8 ul Hexane Spike, 177 mL Chamber)

Test No.	RESPONSE (PPM) No. of Strokes			
	10	15	20	25
1	1500	1500-2000	--	--
2	1000-1500	1000-1500	1500	1500-2000
3	1500	1500-2000	2000	2000-2500
4	1500	1500-2000	2000-2500	2500-3000

		TABLE B-4. PRELIMINARY LABORATORY RESULTS	
Chemical	Material	Chemical Exposure Time (min)	DT Response (ppm)
Acetone	mdl		100
	Viton/Chlorobutyl	30	100-1000
		30	100-1000
		30	
		background	<100
	PVC	15	1000-3000
		15	1000-3000
		15	3000-6000
		background	100-1000
Methylene Chloride	mdl		100
	Viton/Chlorobutyl	15	300-500
		15	C100
		15	300-500
		background	<100
	PVC	15	1000-1500
		15	100-300
		15	1000-1500
		background	<100
	Chemrel Max	15	leak
		15	<100
		15	<100
		background	<100
Ethyl Acetate	mdl		200
	Viton/Chlorobutyl	15	200 (@ 22 hr)*
		15	<200 (@ 90 hr)
		15	<200 (@ 104 hr)
		background	<200
Hexane	mdl		50
	PVC	15	50-150
		15	<50
		15	50-150
		background	40

(continued)

TABLE B-4. PRELIMINARY LABORATORY RESULTS

Chemical	Material	Chemical Exposure Time (min)	DT Response (ppm)
Methanol	mdl		50
	Viton/Chlorobutyl	15	50-100
		15	<50
		15	<50
		background	<50
	PVC	15	<50
		15	<50
		15	<50
		background	<50
Toluene	mdl		50
	Chemrel Max	15	<50
		15	<50
		15	<50
		background	<50
Gasoline	mdl		100
	Viton/Chlorobutyl	0.5	300-500
		0.5	300-500
		0.5	>2500
		background	<100

DT   Detector Tube
mdl   Minimum Detectable Limit. mdl is based on the first indicating line for the DT's.
*   ug extracted (@ delay time between exposure/decon and analysis) all others analyzed at
            approximately 21 hour delay time.

APPENDIX C

DTS CALIBRATION AND TEST DATA

TABLE C-l. DTS TEST RESULTS -- PVC/ACETONE

Calibration Data			Regression Output	
ppm	Area		Constant:	290776
666	1108415		Std. Err. of Y Est.:	218479
2250	3743744		R Squared:	0.9999
7946	1E+07		No. of Obs.:	4
86004	1.2E+08		Degrees of Freedom:	2
			X Coeff.:	1423470
			Std. Err. of Coef.:	3055

Test Data			
No.	Area	ug in sample	ug in whole swatch
1	1.43E+09	1004	19839
2	2.60E+09	1826	36082
3	2.30E+09	1616	31932

Calibration Curve

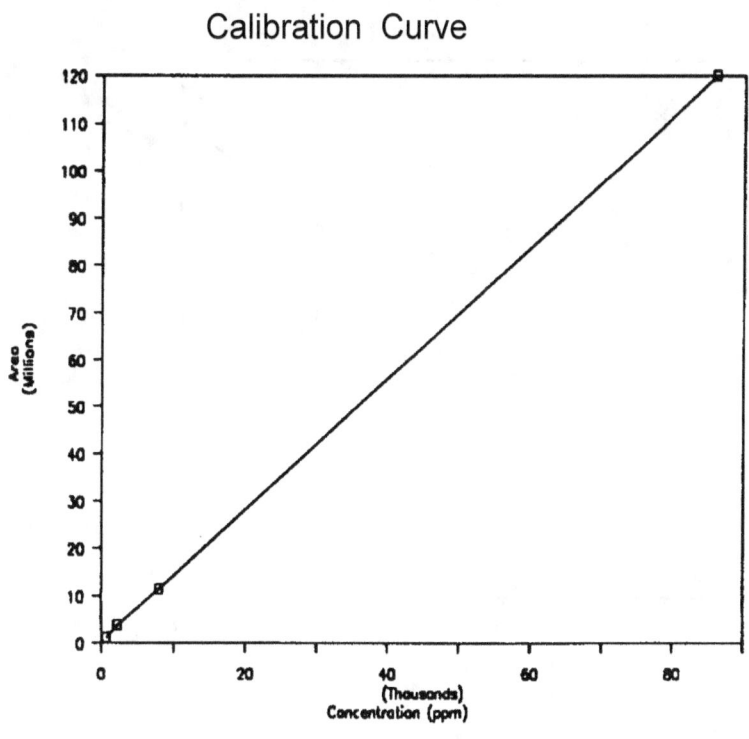

41

TABLE C-2. DTS TEST RESULTS -- VITON/CHLOROBUTYL/GASOLINE

Calibration Data		Regression Output	
ug	Area	Constant:	-7.2E+07
80	2.3E+08	Std. Err. of Y Est.:	2E+07
160	5.0E+08	R Squared:	0.99724
240	8.2E+08	No. of Obs.:	3
		Degrees of Freedom:	1
		X Coeff.:	3693437
		Std. Err. of Coef.:	194387

Test Data

No.	Area	ug in sample	ug in whole swatch
1	2.10E+08	75	1485
2	1.90E+08	70	1388
3	2.30E+08	81	1605

Calibration Curve

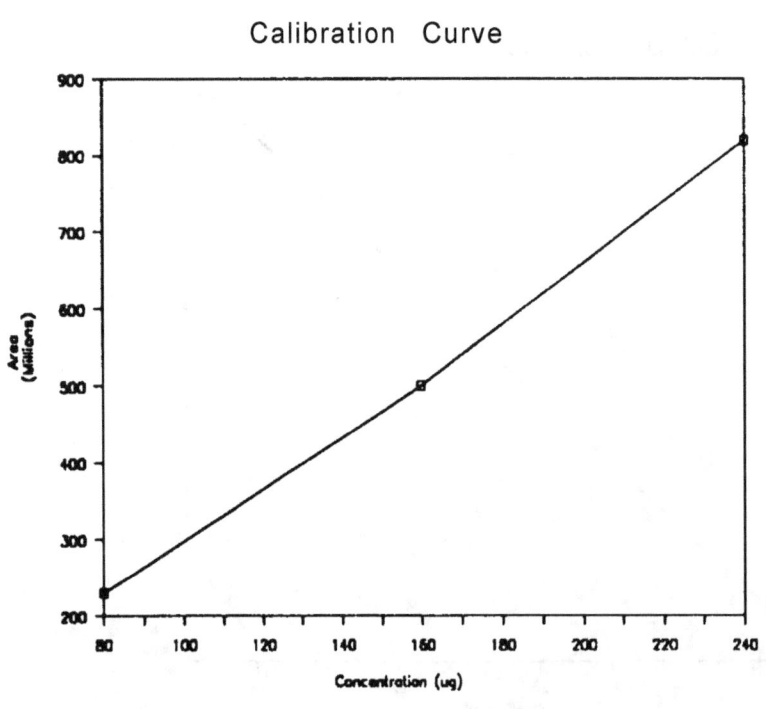

TABLE C-3. DTS TEST RESULTS -- VITON/CHLOROBUTYL/METHANOL

Calibration Data		Regression Output	
ug	**Area**	Constant:	-4281
135	264759	Std. Err. of Y Est.:	180212
475	402164	R Squared:	0.9999
2201	2538400	No. of Obs.:	3
		Degrees of Freedom:	1
		x Coeff.:	1144773
		Std. Err. of Coef.:	115001

Test Data			
No.	**Area**	**ug in sample**	**ug in whole swatch**
1	3.90E+07	34	672
2	2.78E+06	2.4	47
3	1.57E+06	1.4	28

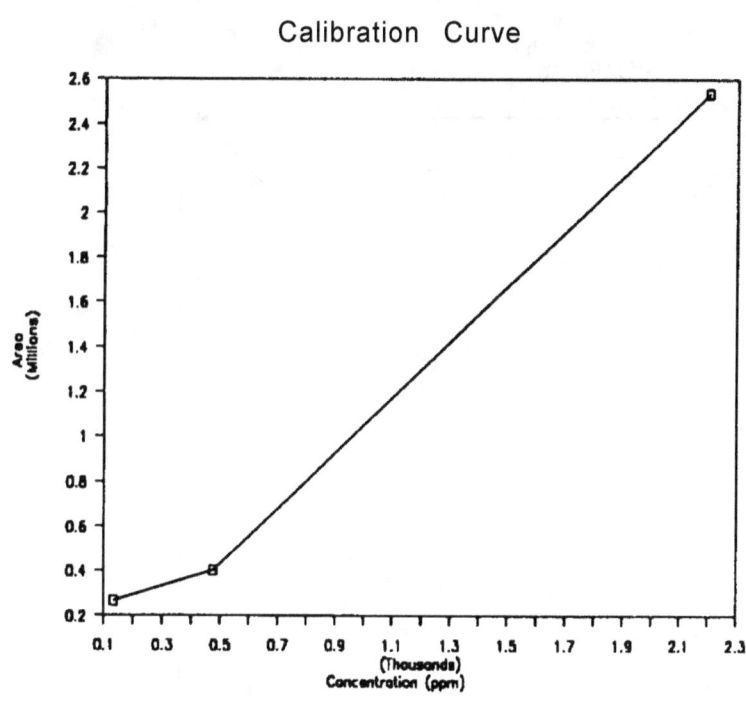

Calibration Curve

43

TABLE C-4. DTS TEST RESULTS - VITON/CHLOROBUTYL/ METHYLENE CHLORIDE

Calibration Data		Regression Output	
ug	Area	Constant:	1.8E+07
393	1.8E+08	Std. Err. of Y Est.:	2478195
524	2.4E+08	R Squared:	0.9999
786	3.8E+08	No. of Obs.:	3
		Degrees of Freedom:	1
		X Coeff.:	503122
		Std. Err. of Coef.:	8757

Test Data

No.	Area	ug in sample	ug in whole swatch
1	1.39E+07	63	1253
2	1.86E+07	73	1437
3	2.84E+07	92	1821

Calibration Curve

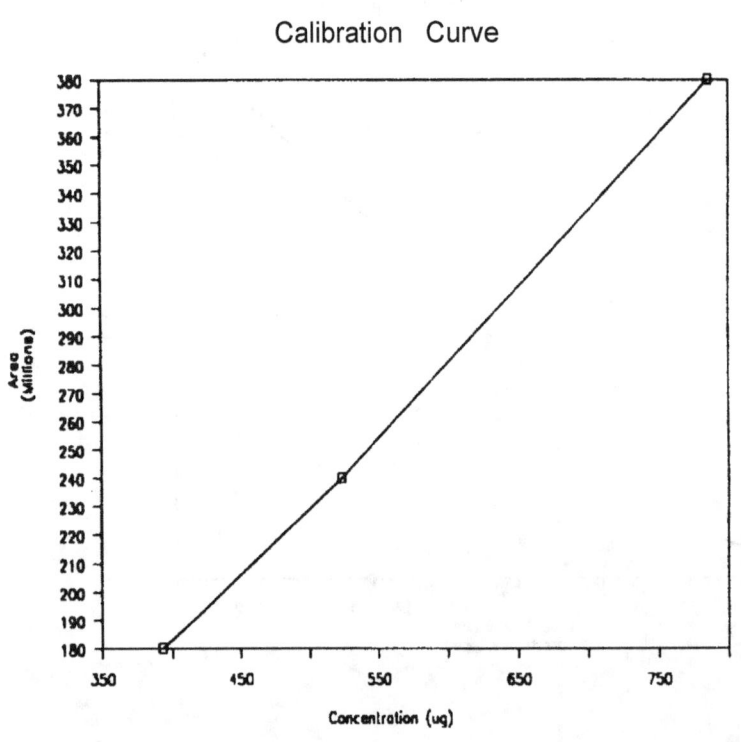

APPENDIX D

DRAFT PROTOCOL FOR FIELD EVALUATION

NONDESTRUCTIVE TESTING AND FIELD EVALUATION OF CHEMICAL PROTECTIVE CLOTHING

FEMA CONTRACT NO. EMW-89-C-3045

TASK 4 - DRAFT PROTOCOL FOR FIELD EVALUATION

OCTOBER 1990

The objective of this task is to evaluate a technique designed to determine the presence of chemical contamination in chemical protective clothing (CPC). Of interest is the potential effectiveness, usefulness, and acceptability of the technique.

EQUIPMENT

1.   2, Saranex-Tyvek coveralls (attached booties), gloves (nitrile), duct tape.

2.   Appropriate protective equipment for participants: Saranex-Tyvek coveralls, (3-each), gloves (natural overglove with Silver Shield underglove), overboots (neoprene), SCBAs, safety glasses, extra natural rubber gloves, etc.

3.   -10 gallons ethyl acetate

4.   Mannequin

5.   3-large decon pools, l-small hand decon tube, buckets, long-handled brushes, decon agent (Unscented Liquid Tide), polyethylene sheeting for ground cover, paper towels, hose, chairs, water cooler and cups, waste buckets, clipboards (3), stopwatch, etc.

6.   Applicators for decon water and chemical contamination (e.g., sprayers).

7.   Drying rack (two ring stands with cross bar and 10 alligator clip hangers).

8.   6, 6-oz volatilization chambers, hotplates, beakers, stir bars, and 1, 2-channel thermometers.

9.   2 Drager bellows pumps, with counters.

10.   Detector tubes CH20201 (Ethyl Acetate}, four boxes.

11.   Specimen dies, Teflon tape, release paper, plastic bags, hammer, specimen pad, etc.

12.   Test materials: neoprene, polyvinyl chloride, viton/chlorobutyl.

Arthur D. Little will be responsible for supplying all items except SCBAs and extra air bottles. Each participants will need to bring three 4500# bottles.

PROTOCOL

The testing protocol will include contamination of splash suits under simulated conditions, decontamination of the suits, and subsequent analysis of the suits for residual contamination. The study will include three materials and one chemical challenge. A minimum of two participants will be needed.

The participants will be asked to wear appropriate protective clothing and respiratory protection while participating in this study. The participants will also be asked to sign a consent form. The study will be conducted at ADL in Cambridge. The tentative date for the study is mid-October.

TEST MATRIX - NUMBER OF ANALYSIS PER CHEMICAL/EXPOSURE CONDITION

Materials	Virgin Material	Exposed
Neoprene	3	5
PVC	3	5
Viton/Chlorobutyl	3	5

Contamination/Decontamination

Participants will wear the following during the contamination, decontamination, and contamination containment portions of the field study: SCBA, Saranex-Tyvek coverall, natural rubber overgloves with Silver Shield™ undergloves, and neoprene boots.

1.   Layout and prepare contamination and decontamination stations according to Figure 1. Fill all appropriate containers with water, decon agent, etc.

2.   Suit-up mannequin in protective clothing; tape glove/suit and boot/suit interfaces. Apply tape tabs to suit (see Figure 2). Orient tabs such that the folded end of the tape faces down,

3.   Place mannequin in metal contamination trough. Fill applicator with ethyl acetate, and spray mannequin with challenge chemical for 7.5 minutes (front only).

4.   Remove manrequin from contamination pool and place in decontamination pool.

5.   Rinse mannequin with water for 5 minutes.

6.   Wash mannequin with decon agent and brushes for 5 minutes.

7.   Move mannequin to final rinse pool.

8 . Rinse mannequin with water for 5 minutes.

9 . Remove tape tabs for analysis.

10. Set mannequin aside to air dry.

11. Return to exposure area for transfer of spent chemical. Use the siphon pump to transfer the ethyl acetate remaining in the exposure pool to a polypropylene bucket. After transfer, secure a lid on the bucket and set aside. Proceed to personal decontamination and equipment drop.

Analysis

Appropriate eye and hand protection will be worn during the analysis portion of the field study.

1.   Pat tape tabs dry with paper towel and remove material swatches.

2.   Hang swatches on drying rack for 1 hour.

3.   Place stir bars and swatches in volatilization chambers. Purge the inlet and outline tubes of the chambers using the bellows pump, assemble chambers, and **CLOSE VALVES.**

4.   Fill beakers half way with water and place on hot plate.

5.   Place chambers in beakers and turn on heat and stirrer.

6.   Extract for 1 hour refilling beaker with hot water as needed.

7.   After 1 hour, prepare a detector tube for analysis. Break off the tips of the tube and fit it into the bellows pump. Attach the free end of the tube to the outlet hose of a chamber. *(Note: The arrow on the detector tube should face the bellows pump.)*.

8.   Depress and release the bellows pump 1 time (stroke #1). Wait until chain is taunt.

9.   Depress and release the bellows pump a second time (stroke #2). Open the outlet valve.

10.   Open the inlet valve.

11.   Continue to depress and release bellows pump for the number of strokes printed on the detector tube (20).

12.   Remove detector tube from bellows pump and outlet tubing. Inspect tube for color change and record on data sheet. Report tube responses as a range of full indication lines.

13.   Remove chamber from beaker. (Caution: BEAKER WILL BE HOT)

14.   Wipe chamber with paper towel, open, and remove material swatch. Set chamber aside for next analysis.

S c h e d u l e

Day 1 -

8:30-9:30 am	•	introduction and set-up
9:30 am	•	donning of eye protection
9:30- 10:00 am	•	background analysis
lO:OO- 12:30 am	•	virgin material analyses (Neoprene, PVC, Viton/Chlorobutyl)
10:30- 12:30 am	•	dry run exposure
12:30-1:30 pm	•	lunch
1:30- 1:45 pm	•	donning of protective equipment
1:45-2:15 pm	•	Neoprene exposure/decon
2:15-2:30 pm	•	clean-up
2:30-3:00 pm	•	decon and doffing of protective equipment
3:00-3:15 pm	•	water break
3:15 pm	•	donning of eye and hand protection
3:00-4:30 pm	•	Neoprene analysis
4: 15-4:30 pm	•	background analysis
4:30-5:00 pm	•	clean-up

Day 2 -

8:30-930 am	•	set-up
9:00-9:30 am	•	donning of protective equipment
9:30- 10:00 am	•	PVC exposure/decon
10:00-10:15 am	•	clean-up
10:15-10:30 am	•	decon and doffing of protective equipment
10:30-10:45 am	•	water break
10:45 am	•	donning of eye and hand protection
10:45- 12:00 am	•	PVC analysis
11:15-12:00 am	•	background analysis
12:00-1:OO pm	•	lunch
1:OO- 1:30 pm	•	donning of protective equipment
1:30-2:00 pm	•	Viton/Chlorobutyl exposure/decon
2:00-2:15 pm	•	clean-up
2:15-2:45 pm	•	decon and doffing of protective equipment
2:45-3:00 pm	•	water break
3:00 pm	•	donning of eye and hand protection
3:00-4:15 pm	•	Viton/Chlorobutyl analysis
3:15-4:00 pm	•	background analysis
4:15-5:00 pm		clean-up & discussion

FIELD DATA SHEET

FEMA DECONTAMINATION STUDY

Name: _____   Date: _____

Temp: _____

Material: _____

Tab Location: _____

Chemical Challenge: _____

Extraction Time: _____

Detector Tube: _____

No. of Strokes: _____

Tube Response (Range): _____

Min. Indication of Tube: _____

Observations: _____

_____

_____

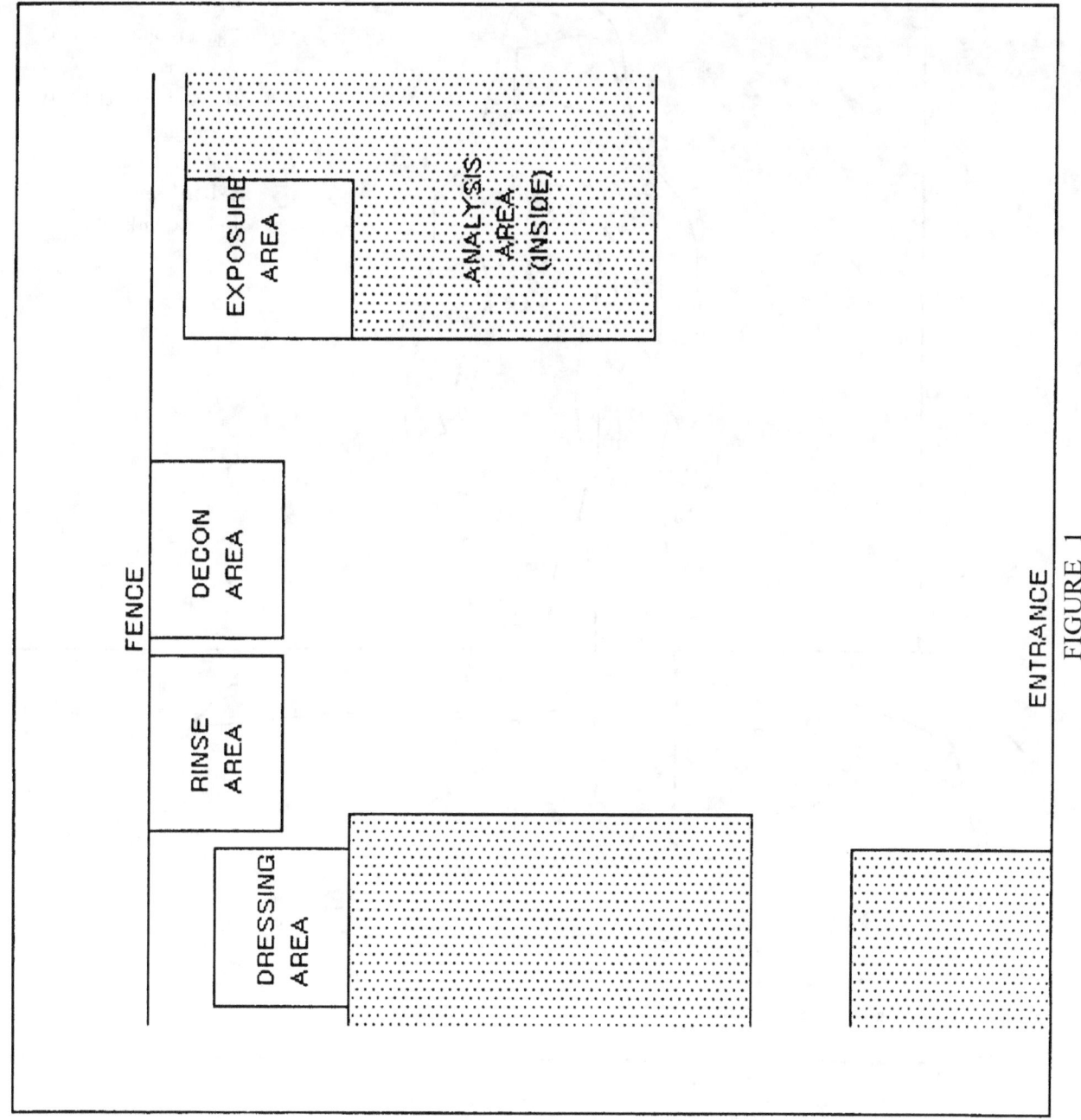

FENCE

EXPOSURE AREA

ANALYSIS AREA (INSIDE)

DECON AREA

RINSE AREA

DRESSING AREA

ENTRANCE

FIGURE 1

53

FIGURE 2. FIELD STUDY MANIKIN

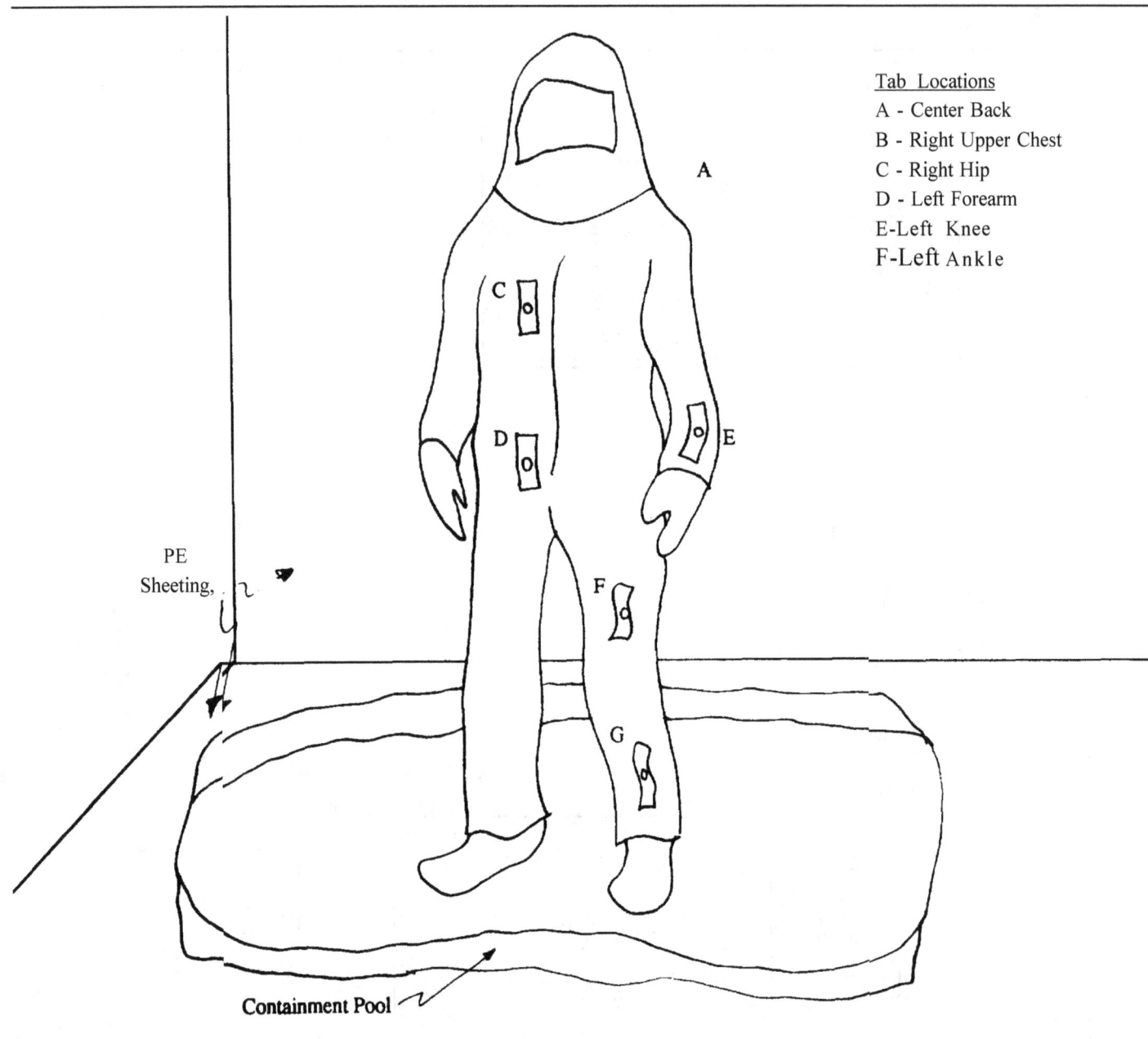

Tab Locations
A - Center Back
B - Right Upper Chest
C - Right Hip
D - Left Forearm
E-Left Knee
F-Left Ankle

APPENDIX E
STEVE STORMENT FIELD STUDY SUMMARY

NON-DESTRUCTIVE TESTING AND FIELD EVALUATION
OF CHEMICAL PROTECTIVE CLOTHING

FINAL REPORT

by:

Steve Storment, Deputy Chief
City of Phoenix Fire Department
Special Operations Section

TECHNICAL

Project Objective --

The objective of this contract is to develop and validate a cost effective means by which the fire department and other hazardous material response team can make decisions on the use/reuse of chemical clothing that has been exposed to hazardous chemicals during a hazardous material incident.

MODIFIER

As a result of this technique to qualitatively measure contamination, application to other fabrics have been demonstrated, the flexibility of this testing technique allowed the participants to qualitatively measure contamination of NOMEX III and leather goods on fire fighters protective ensembles without modification to the testing technique or test apparatus.

BACKGROUND

The evaluation for contamination in protective clothing ensembles has been an on going problem since the awareness of health related illness to contaminated clothing was first suggested in the 1960s with Asbestos workers. Since that time, heightened awareness of chemical protective ensembles, rubbers, and compound materials, have increased in respect to their ability to be decontaminated. Although many different methods exist to test for contamination, they are all of a destructive nature. Until this study, no practical testing technique has been developed that has field use application.

DISCUSSION

The non-destructive testing and field evaluation of chemical protective clothing technique has been achieved. The testing technique used a swatch sampling scheme. The method combines enclosure and attachment requirements. Swatches of CPC are made to face out of a window to allow for contamination and decontamination. Once this Band-Aid[TM] type tab is attached to the garment by the adhesive action of the tape, it becomes an intimate part of the garment with the same exposure possibilities as any other surface area on the protective ensemble.

The swatch of material can be easily obtained when the purchaser obtains chemical protective clothing during their normal purchasing practices. These bolts of materials can be specified to represent the "lot material" that the protective ensemble was fabricated from.

This study used two highly motivated professional fire fighters from the Cambridge, Massachusetts Fire Department Hazardous Materials Team. In the course of one hour's time, a demonstration of the testing method was accomplished. This short time period is not only a verification of the ease of the testing method, but the skill and motivation of the two participants.

COST

The cost of test development was moderate. The benefit to Fire Departments and chemical response personnel is high. The expense of materials used in the test technique would be within the reach of most, if not all, Fire Departments, industrial brigades, and private contractors, Given the average cost of 8 chemical protective ensembles with an average cost of $2,000 (low), $16,000 total, most any fire department or other chemical response group would be willing to invest 5- 10% ($800 - $1,000) in testing equipment to save a $16,000 chemical protective clothing investment.

RECOMMENDATION

The worth of this project is beyond reproach. The U.S. Fire Administration has again demonstrated its forward thinking in the entire area of hazardous material response and chemical protective ensembles.

- Continue refinement in the area of testing equipment. Develop a piece of testing apparatus that would be as portable as field user friendly as the chemical permeation test kit (currently being marketed by Arthur D. Little).

- Fund for development and field testing another method using magnetic enclosures as heat exchange unit. This type of method and apparatus could allow a user to test any piece of protection, on any surface area he/she desires.

- Fund a study that would focus the use of this currently developed test method of the electromagnetic closure system around fire fighter protective clothing. Given the current cost of $400-600 per ensemble, it would make sense to investigate the current contamination of this ensemble. This area in my opinion has the greatest potential for lessening the chemical exposure of fire fighter to low levels of possible cancer causing substances found in common structure fires. Using "Project Firesmoke," funded by NIOSH and the U.S. Fire Administration, the data from that study could be the impetus to drive such a study.

- Finally, develop a decontamination strategy based on the testing method. This to me is the logical conclusion to this type of study. This decontamination strategy could be adopted by the Fire Service so we, as an industry, could finally standardize our method of what is dirty and what is clean.

☆ U.S.GOVERMENT PRINTING OFFICE 1991-526-661/40892